PETROGRAPHIC AND CORRELATION STUDY OF THE SAN JOSÉ DE LAS MALEZAS QUARTZ-GOLD DEPOSIT IN SANTA CLARA, VILLA CLARA, CUBA.

Ricardo A. Valls, P. Geo., M. Sc.

Copyright © 2016 Ricardo A Valls

All rights reserved. This publication is protected by copyright and permission should be obtained from Valls Geoconsultant prior to any prohibited reproduction, storage in a retrieval system, or transmission in any form or by any means, electronic, mechanical, photocopying, recording, or likewise. For information regarding permission write to Valls Geoconsultant at 1008-299 Glenlake Ave., Toronto, Ontario, Canada, E-mail: vallsvg@gmail.com.

ISBN-13: 978-1533521545
ISBN-10: 1533521549

Contents

Summary 4
Introduction 4
Regional Geology 5
Sample Preparation and Spectral Analysis 8
Petrography 9
Conditions of Formation of these Petrologic Groups 15
    Mafic group 15
    Felsic group 15
Correlation Presentations 16
The Formation of the San José de las Malezas Deposit 20
Dynamic of the Hydrothermal Fluid 24
Discussion 28
Conclusions and Recommendations 29
Acknowledgment 31
References 32
About the Author 35

Summary

A series of nine full petrological descriptions were made from the main host rocks at San José de Las Malezas quartz-gold deposit in Santa Clara, Villa Clara, Cuba. These main petrological groups include leucocratic gabbros, "fresh" (less altered) massive serpentinites, quartz veins and altered (iron and carbonatic alterations) serpentinites. The mafic rocks contain chiefly plagioclase and clinopyroxene as the principal components, with less frequently olivine and a relative large variety of secondary and accessory minerals. Felsic rocks are composed mainly by quartz with different accessory minerals.

The inductively coupled plasma optical emission spectroscopy (ICP-EOS) of the global composition of these two main petrological groups allowed the determination of their characteristic correlation coefficients. This study suggests that gold was contributed not by the hydrothermal process itself, but by the remobilization of this element from the host rocks.

The presence of zeolites within the quartz bodies is interpreted as an indicator of their low-temperature character. It is proposed that quartz veins were formed as a result of the listwaenitization of the serpentinites, combined with the hydrothermal-metasomatic fluids that altered the host rocks after their obduction onto the surface.

Introduction

The main purpose of this investigation was to present a summary of the petrographic characteristics from the San José de Las Malezas quartz-gold deposit, and to show the validity of the use of correlation representations to study the geochemical data and to learn more about the genesis of the outcrop. Although the deposit is known since 1881, and have been studied by several geologist since 1963 (Hantichak 1963; Grushdev et al. 1966; Kanchev et al. 1973; Cabrera and Tolkunov 1979; Arcial and Romero 1985; Cruz and Valls 1989; et al.) this kind of petrogeochemical study has not been done to the present. Samples were obtained during a detailed mapping of the San José de Las Malezas deposit and adjacent areas (Cruz and Valls 1989).

Regional Geology

The San José de Las Malezas quartz-gold deposit is located within the Structural-Facial Zone (S.F.Z.) "Zaza", in the province of Villa Clara in Central Cuba (Fig. 1).

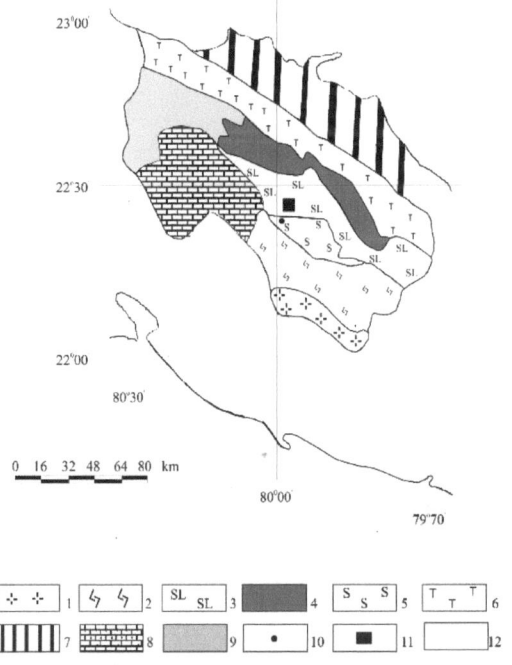

Figure 1. Regional geology of Central Cuba. 1.- Diorites; 2.- Volcanic sequences; 3.- S.F.Z. "Zaza"; 4.- S.F.Z. "Placetas"; 5.- Oceanic crust; 6.- S.F.Z. "Camajuaní"; 7.- S.F.Z. "Remedios"; 8.- Sedimentary formations; 9.- N-Q formations; 10.- City of Santa Clara; 11.- San José de Las Malezas deposit; 12.- Other formations (chiefly metamorphic sequences).

This S.F.Z. is composed of (i) a volcano-sedimentary complex of Lower Cretaceous age (Turonian) located to the south, (ii) the Ochoa Formation composed of limestones and marls of the Eocene in discordant contact to the north, and (iii) the Zurrapandilla Suite, composed of diabasic porphyries, spilites, gabbro diabase, gabbro diabase porphyries, and other gabbroic rocks that cut both described complexes (Cabrera and Tolkunov, 1979).

As representatives of the S.F.Z. "Placetas" we have the Constancia Formation (J_3t), the Veloz Formation (J_3t-K_1b), the Fidencia Formation (K_1b), the Carmita Formation (K_2al-cm), and the Vega Alta Formation (Palaeocene). They are mainly carbonates, with minor conglomerates, sandstones and also siliceous minerals.

Within the S.F.Z. "Zaza" and in tectonic contact with the volcano-sedimentary complex just described, we also find serpentinitic bodies, that form a large massif with an east-west orientation. These intrusions form part of the Cuban hyperbasitic belt of the Upper Cretaceous, and they are commonly interpreted as the remains of an ancient oceanic crust.

The serpentinites are massive, fractured, light green rocks, with a reticular structure due to the uneven distribution of chrysolite and antigorite in the rock, and the presence of magnetite, chromite and spinel. They are frequently cut by diabase, micro diabase and porphyritic diabase dikes from the Zurrapandilla Suite.

GEOLOGY OF THE DEPOSIT

In the vicinity of this deposit we find serpentinites, gabbroic rocks, diabase dikes, diorites, and quartz veins (Fig. 2).

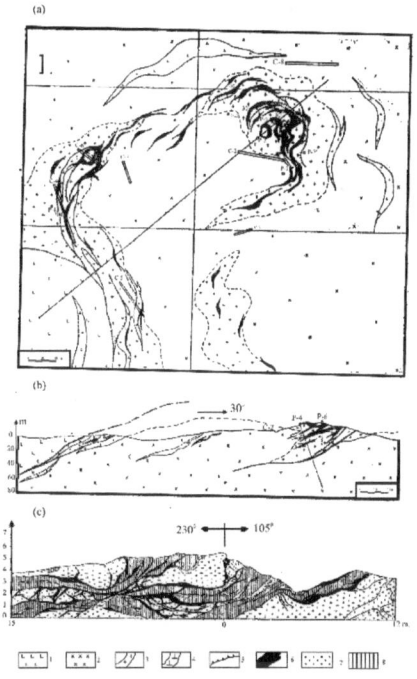

Figure 2. Geology of the San José de Las Malezas quartz-gold deposit, in Santa Clara, Cuba (From Cabrera and Talkunov, 1979). (a)- Local geology; (b)- Geological cut; (c)- Wall of the open pit. 1.- Serpentinites; 2.- Diorite porphyry; 3.- Dike of porphiritic diorites; 4.- Dike of porphyritic diabases; 5.- Tectonic faults; 6.- Ore bodies; 7.- Altered rocks; 8.- Quartz veins.

Serpentinites crop out to the west. At the centre of the area we find a big intrusion of gabbros, represented by grey, fine to medium grained rocks with a gabbroic texture. Both the gabbros and the serpentinites are cut by diabase, micro diabase and porphyritic diabase dikes.

These are dark grey compact rocks, fine grained to microgranular, with ophitic and microphytic structures.

We also find quartz veins that vary from 10 to 100 meters in length and 0,2 to 5 meters in width. The morphology of these quartz bodies is very complicated. Most frequently they form small lens, with abundant apophasis, wedges, and stockwork-like structures, intercrossing with subordinate diagonal veins. They also form sinusoidal structures like "waves". Post-ore tectonic movements added more complications to the structure of these formations. The more abundant quartz is a milky white variety, that frequently turns reddish because of the presence of hematite.

Sometimes one can observe the presence of porous quartz, due to the weathering, with the pores filled with limonite and other iron minerals. Copper mineralization occurs where the quartz veins are in contact with the fractured host rocks, and forms pockets, nests, crusts, and impregnations of altered minerals, containing an uneven distribution of gold.

Sample Preparation and Spectral Analysis

Several grab samples were taken from the deposit for petrographic descriptions of the host rocks. Petrographic study of these samples was done by Mauro Ordoñez Morejón from the Central Laboratory of the Geological Enterprise "Santa Clara". The geochemical data for this study was obtained from the spectral analysis of channels samples, taken by the author and his college Jorge Cruz, in 1986, from a cut perpendicular to the deposit's structures. The volume of each channel sample was of 3 x 3 x 100 cm^3, with an approximated weigh of 2 to 3 kg each.

Before sampling, the surface was scraped to avoid as much as possible the effect of contamination. These samples were also analysed at the Central Laboratory of the Geological Enterprise "Santa Clara" by Santa García Gonzalez and her Bulgarian adviser Lazarina Lazarova Ignátova. They used an ICP-EOS with a dispersion camera, and an internal control of 30% of the total amount of analysed samples. Both, the field control (15%) and the internal control were satisfactory. The maximum error detected was of 20% for gold determinations. This method gave information on As, Ag, Ni, Cu, Co, Zn, Ga, Mn, and Au.

Petrography

Composition and other notes on the analysed specimens are given in Table 1. Plagioclase composition was determinate by measurement of the angle of extinction and is expressed in mole% of Anorthite. A more detailed description of specimens from Table 1 follows.

Table 1. Composition of the analysed samples from San José de Las Malezas quartz-gold deposit.

Composition	II-10	II-9	I-2	I-3	I-1	II-7	I-6	I-5	III-1
Quartz							P » 100%	P	
Plagioclase	P >50%	P >50%	P 50-60%	P 45%	P		A		P 50%
Clinopyroxene	P 40%	P 20-25%	P 20-25%	P 25%					P 40-45%
Orthopyroxene						A			
Olivine	P 3%								
Sphene									P >> 2%
Amphibolites	S	S		S	S				
Zeolites	S	S	S	S	P		A		S
Chlorite		S	S	S	P		A	P	
Serpentinite	S		S			P > 95%			
Calcite				S					
Magnetite	A	Tr	A	Tr	Tr	A	A	Tr	Tr
Spinel						A			
Limonite	A	Tr	Tr	Tr	Tr	A	A	A-S	Tr
Name	Olivinic gabbro	Amphibolitic gabbro	Oriented gabbro	Uralitized gabbro	Cataclastic gabbro	Serpentinite	Hydrothermal quartz	Altered quartz zone	Micro-dolerite

P- Principal (> 3%)   A- Accessory (< 1%)
S- Secondary (> 1%)   Tr- Not common.

Sample II-10 (Slightly Serpentinitic Olivine Gabbro)

This is a grey, fine grained, weather brown rock, fractured to some degree.

It presents a quasi-oriented gabbroic texture. Main minerals are plagioclase (50%), clinopyroxene (40%) and olivine (3%). As secondary minerals we have serpentine, prehnite, and actinolite. Plagioclase constitutes the majority of the rock and it is represented by labradorite/bitownite (An 70%).

They form prismatic crystals that are more or less oriented in the direction of their larger borders. Plagioclase are relatively fresh, fractured and with crystals ranging in size from 0.46 to 3.22 mm. Clinopyroxene crystals (diopside) range in size from 0.36 to 2.30 mm and they are frequently altered. Because of that, some crystals surrounded by actinolite can be found. Olivine forms irregular grains, very fractured and altered to serpentinite (light green) and iddingsite (yellow). The metallic component is represented by magnetite and limonite.

Sample II-9 (Amphibolitic gabbro)

Grey, fine grained rock with very thin veinlets of white zeolite. The sample shows an oriented gabbroic texture. Main minerals are plagioclase (> 50%) and clinopyroxene (20-25%). Secondary minerals form almost 20% of the sample, and are represented by actinolite, uralite, chlorite, albite and zeolite. Plagioclase constitutes most of the rock an is represented by a labradorite with An 56%. Crystals are prismatic and macled. They are usually altered to albite plus zeolite plus chlorite, but in a lesser degree by comparison with the clinopyroxenes. These crystals have a maximum size of 4.5 mm and their surfaces are usually corroded.

Clinopyroxene is represented by altered xenomorphic to hipidiomorphic crystals of diopside. Sometimes they are completely substituted by the amphibole (uralite and actinolite).

Sample I-2 (Oriented Gabbro)

Greenish, altered rock with phenocristals of clinopyroxene, and a gabbroic texture. Main minerals are plagioclase (>50-60%) and amphibolitized clinopyroxene (20-25%). Secondary minerals are represented by zeolite, chlorite, serpentine, and sericite for a 10 to 15% of the global composition and almost a 1% of magnetite and limonite. The labradorite (An 60%) is relatively fresh compared to the clinopyroxenes and form elongated prisms up to 2.3 mm in length. Plagioclase is slightly oriented. Alteration is represented by zeolite and in a lesser degree by thin needles of actinolite. Clinopyroxene also forms elongated prisms with sizes of up to 5 mm. This elongated habitus could be an indication of a rapid cooling of the magma (Lofgren ,1980).

Alteration is represented by the formation of amphibolites plus serpentine plus sericite, chlorite and small inclusions of plagioclases. In some clinopyroxenes we can observe the following progressive substitutions:

(a)- Diopside ---> Actinolite ---> Chlorite

(b)- Diopside ---> Serpentinite ---> Actinolite ---> Chlorite ---> Mica

Fractured are filled by altered minerals, mainly zeolite. Magnetite and limonite are present everywhere as small grains and metallic dust.

Sample I-3 (Amphibolitic Gabbro)

Grey, fine grained rock with a gabbroic texture. Main minerals are plagioclase (> 45%) and amphibolitic clinopyroxene (25%). As secondary minerals we have actinolite, chlorite, calcite, and zeolite forming 25-30% of the global composition of the rock. The surface of the labradorite (An 60%) is corroded, but in general they are less altered than the clinopyroxene (diopside). The most common altered minerals present are epidote-clinozoizite and other zeolitic minerals, with a maximum size of 5 mm. Diopside forms sometime crystalline aggregates, but most frequently they appear as isolated prisms, with irregular faces and males. They are intensively altered. In some crystals we can observe a zonal composition of actinolite and uralite in the border to diopside in the core.

Sample I-1 (Cataclastic gabbro)

Greenish, extremely cataclastic rock. By the relict diopsides and plagioclase (labradorite?), we can assume that the initial composition of this rock was similar to those of the gabbros. Almost all the rock is composed by scales of zeolitic minerals (zoizite, clinozoizite, etc.), prehnite, chlorite, and albite. All these minerals are altering the plagioclase. The diopside appears also completely altered. Uralite and actinolite can substitute the diopside by pseudomorphism and sometimes they completely replace the clinopyroxene.

Sample II-7 (Lizardite serpentine)

Massive, purple-green rock, with reticular texture. Main mineral is lizardite (> 95%). We also find orthopyroxene and bastite as the result of the alteration of the hypersthene. There is also chrysotile and magnetite which increase the reticular character of the sample. We find also limonite which gives the rock its distinctive weather brown colour. Finally, we see isolated and elongated crystals of 0.21 mm of a brownish spinel.

Sample I-6 (Hydrothermal quartz)

The sample is composed almost completely by crystalline aggregates of allotriomorphic grains of quartz, of up to 5 mm in length. Between these grains we find chlorite (which transforms to mica), epidote and also isolated grains of magnetite of up to 0.36 mm in length sometimes altered to limonite. In some fractures we can find crystals of K-feldspar (microcline) of up to 0.36 mm in length.

Sample I-5 (Quartz-chlorite rock from the altered zone)

Hydrothermal rock composed mainly by fine fibrous aggregates of chlorite, sometimes impregnate with limonite and granoblastic aggregates of quartz. The sample is very fractured with limonite filling almost all the fractures.

Sample III-1 (Micro dolerite)

Dark, very fine grained rock with ophitic texture. Main minerals are plagioclase (> 50%), clinopyroxene (40-45%) and sphene (> 2%). As secondary minerals we have zeolite, epidote and albite (?). Plagioclase is bytownite (An 71%). It is usually twined and 0.48 mm in length. They constitute the majority of the rock. In some parts they are altered to albite and epidote. Clinopyroxene is augite, and it is almost as abundant as the bytownite, but it locates itself in the free spaces between the plagioclase crystals. They are no more than 0.36 mm in length. Sphene is relatively abundant in the sample. It appears disseminated in form of small crystalline aggregates of 0.01 to 0.06 mm in length. They sometimes are altered to leucoxene.

Conditions of Formation of these Petrologic Groups

Mafic group

As detailed earlier the typical mineral assemblage of these gabbros include chlorite + albite + zoisite + calcite + magnetite + titanite (sphene). The coexistence of zeolites and amphibolites indicates that most of the pre-existing prehnite was brokedown to zoisite. Spear (1993) suggests reaction (1) to explain this process. Actinolite was most probably formed via reaction (2).

(1) Prehnite = zoisite + grossular + quartz + H_2O

(2) Low-Al chlorite + calcite + oxides = actinolite ± zoisite + CO_2 + H_2O

Reactions (1) & (2) give rise to the diagnostic **Greenschist facies assemblage** chlorite + albite + zoisite + actinolite + quartz ± calcite ± titanite. This points out to moderate pressures and temperatures of regional metamorphism (Yardley 1989).

Felsic group

Although the composition of these quartz veins is almost 100% pure SiO_2, they have a wide range of secondary and accessory minerals which include zoisite, clinozoisite, magnetite, K-feldspar (microcline), micas, and hematite. This assemblage, and chiefly the presence of several zeolitic minerals, suggests low temperatures and low pressures conditions of formation of the San José de Las Malezas deposit. Therefore, it is suggested that the formation of these quartz veins (as the final result of a listwaenitic process), did occur after the greenschist facies of regional metamorphism that affected the host rocks.

Correlation Presentations

The graphic representation of correlation analysis allows a better comprehension on how the elements are interrelated and gives us some insights on the genesis of the deposit and other important information. In 1988, a method for the graphic representations of the correlation analysis was developed by the author and was published under the name of "The arc method" (Valls and Nuñez, 1988).

This kind of graphics allow us to determine **the correlation coefficient** of each studied group. Two of these graphic constructions are given in Figures 3 and 4. One begins by arranging all the correlations from positive to negative values, in a decreasing order. After that we need to transform those values to a variable of length. The stronger the correlation, the shorter the distance between the pair of studied elements. I suggest the use of equation (3) for this transformation.

(3) $L = N + (1 - (r_i/r_{max}))$

Where,

r_i- is the value of the correlation for the studied pair.

r_{max}- is the maximum value of correlation in the studied sample.

N- is the minimum distance between the pair of elements with the strongest correlation. If you are planning on using a standard paper (letter), I suggest you use N = 4 cm.

L- Is the distance between correlated elements.

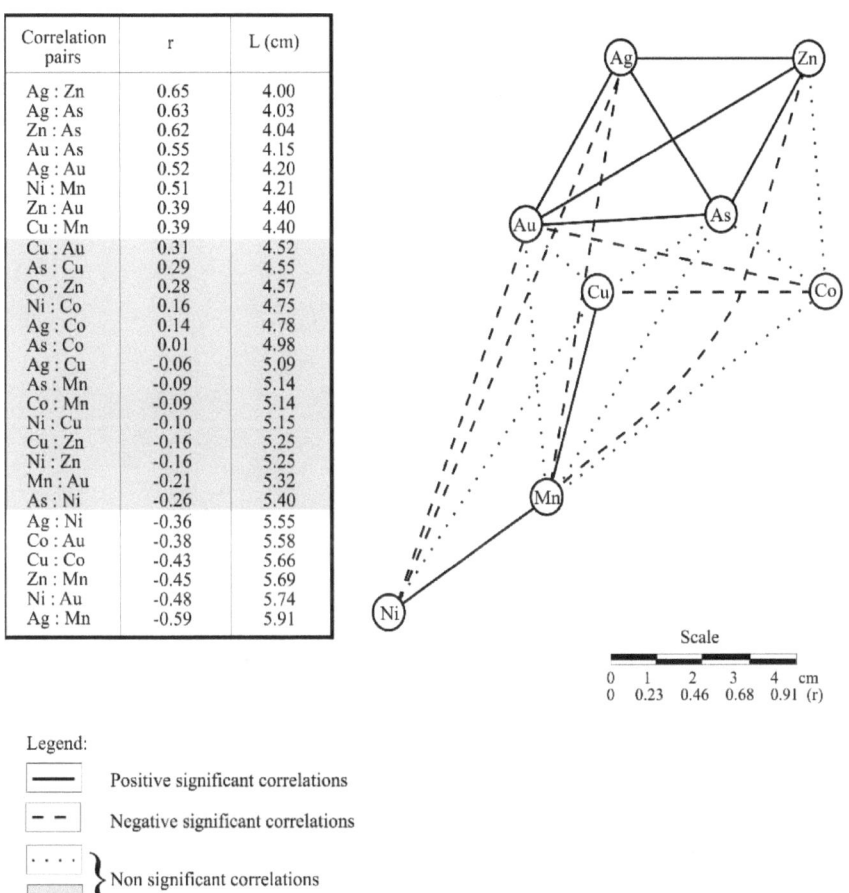

Correlation pairs	r	L (cm)
Ag : Zn	0.65	4.00
Ag : As	0.63	4.03
Zn : As	0.62	4.04
Au : As	0.55	4.15
Ag : Au	0.52	4.20
Ni : Mn	0.51	4.21
Zn : Au	0.39	4.40
Cu : Mn	0.39	4.40
Cu : Au	0.31	4.52
As : Cu	0.29	4.55
Co : Zn	0.28	4.57
Ni : Co	0.16	4.75
Ag : Co	0.14	4.78
As : Co	0.01	4.98
Ag : Cu	-0.06	5.09
As : Mn	-0.09	5.14
Co : Mn	-0.09	5.14
Ni : Cu	-0.10	5.15
Cu : Zn	-0.16	5.25
Ni : Zn	-0.16	5.25
Mn : Au	-0.21	5.32
As : Ni	-0.26	5.40
Ag : Ni	-0.36	5.55
Co : Au	-0.38	5.58
Cu : Co	-0.43	5.66
Zn : Mn	-0.45	5.69
Ni : Au	-0.48	5.74
Ag : Mn	-0.59	5.91

Legend:
— Positive significant correlations
- - Negative significant correlations
⋯ } Non significant correlations

$L = N + (1 - (r_i / r_{max}))$

$N = 4$ cm.

Figure 3. Graphic representation of the correlation analysis for the Mafic group of the San José de Las Malezas quartz-gold deposit. One can see two separated groups, one of rock-forming elements (Ni, Co and Mn) and the other of ore-related elements (Ag, As, Au, Zn and Cu). See text for proposed correlation coefficients.

Correlation pairs	r	L (cm)
As : Cu	0.91	4.00
Ag : Co	0.81	4.11
Cu : Co	0.71	4.22
Co : Zn	0.69	4.24
As : Zn	0.63	4.31
Cu : Zn	0.55	4.40
Ni : Co	0.40	4.56
Co : Au	0.36	4.60
As : Au	0.31	4.66
Ga : Mn	0.31	4.66
Ni : Zn	0.30	4.67
Cu : Au	0.30	4.67
Co : Ga	0.30	4.67
Ni : Au	0.25	4.73
Zn : Au	0.20	4.78
Ni : Ga	0.17	4.81
Ag : Au	0.14	4.85
Zn : Ga	0.08	4.91
Ag : Ga	0.01	4.99
Ni : Mn	-0.01	5.01
Ga : Au	-0.01	5.01
Ag : Ni	-0.02	5.02
As : Ni	-0.06	5.07
Zn : Mn	-0.07	5.08
Ni : Cu	-0.10	5.11
Cu : Ga	-0.11	5.12
Ag : Co	-0.17	5.19

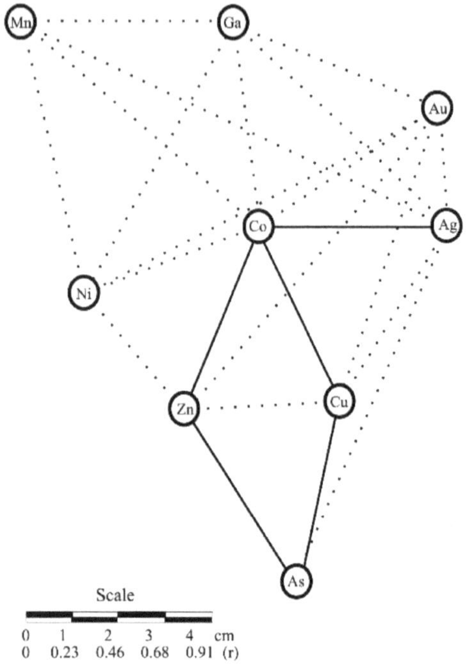

Scale

0   1   2   3   4   cm
0   0.23 0.46 0.68 0.91 (r)

Legend:

——— Positive significant correlations

····· } Non significant correlations
▭ }

$L = N \cdot (1 - (r_i / r_{max}))$

$N = 4$ cm.

Figure 4. Graphic representation of the correlation analysis for the Felsic group of the San José de Las Malezas quartz-gold deposit. Here we also have two groups, one composed by Cu, As, Ag and Zn; and a second composed by Ni, Co and Au. Both groups share correlations with As and Ag, and present a negative sensible correlation with Mn. Notice that Au and Cu are in separated groups. See text for proposed correlation coefficients.

Next, one places the elements with the strongest positive correlation at "N" cm from each other. For example, in the Mafic group represented in Figure 3, that pair will be (As:Au). Next we look for an element that forms a pair with one of those two. The first one is Zn:As (L = 4.1) and the second one is Zn:Au (L = 4.1).

Now take a compass with an opening of 4.1 cm and from the position of the As, and trace a segment of an arc. Do the same thing from the position of Au with the same opening of 4.1 cm. The point where both arcs cut, will decide the position of the element Zn.

After positioning all the elements, we begin to connect them with lines that will reflect the character of the correlation link. For example, positive significant correlations will be represented by a continuous line, while negative significant correlations will be identified by a dash line. Non-significant correlations may be represented by doted lines and you can choose to put them or not in the graph. The objective is to obtain a clear picture of the relative spatial distribution of the elements, and not to represent all the possible links.

Now look at the obtained Figure 3. Is easy to see two separated groups, one of rock-forming elements (Ni, Co and Mn), and a second of ore-related elements (As, Au, Ag, Zn and Cu). Therefore, one can propose two main correlation coefficients (C.C.) for this group:

$C.C._{mafic\ 1} = (As \times Au \times Ag \times Zn \times Cu)$

$C.C._{mafic\ 2} = (Mn \times Ni \times Co)$

The correlation coefficient for the Felsic group is shown in Figure 4. Here we also have two groups, a first one composed by Cu, As, Ag and Zn; and a second by Ni, Co and Au. Both groups share correlations with As and Ag, and present negative sensible correlations with Mn.

One can suggest the following correlation coefficients for this group:

Error!

A complete study on the characteristic correlation coefficients of the San José de Las Malezas deposit can be consulted in a previous paper (Valls 1995a).

The Formation of the

San José de las Malezas Deposit

Our model begins with the protrusion of an ultramafic body of harzburgite composition through a ridge axis. It has been suggested elsewhere (Valls and Gonzalez, 1987), that this ultramafic magma could have been subjected to a partial melting process, due to which we could have obtained an area of gold enrichment by gravitational separation inside a magma chamber.

After the protrusion, the sea water and the heat from the upwelling zone initiated the serpentinization of the rocks. During this process, Mn^{2+} was liberated because of the decomposition of olivine. Another possible mineral that could liberate Mn^{2+} during the serpentinization of these rocks is pyrophenite ($MnTiO_3$) (W. Trzcienski, personal communication).

Under these anaerobic conditions, only divalent manganese minerals could be formed. Two possible contributing reactions are given bellow, (i) the formation of pyrochroite from tephroite (4), and (ii) the formation of rhodochrosite from tephroite by carbonatic sea water (5).

(4) $Mn_2SiO_4 + 2H_2O ---> 2Mn(OH)_2 + SiO_2$

(5) $Mn_2SiO_4 + 2H_2CO_3 ---> 2MnCO_3 + SiO_2 + 2H_2O$

Since pyrochroite is a much less common mineral than rhodochrosite (Crerar et al., 1976), we can assume that the formation of rhodochrosite was the most probable reaction. In fact, when we study the mineralogy of this type of deposit World-wide, we usually find references to the presence of rhodochrosite, rhodonite, pyrolusite, and other manganese minerals (Baranova and Ryzhenko, 1981; Farfel, 1984; Baranova and Koltsov, 1987; Camus, 1990; Rodriguez and Warden, 1993; etc.).

The formation of rhodochrosite will take place for as long as olivine is decomposed during the serpentinization of the rocks in the heated zone near the rift. Since we find relics of olivine crystals in these serpentinites, we can assume that serpentinization was stopped before the obduction onto the colliding continental plate during the final stage of the ocean closure.

After the obduction of these rocks onto an aerobic environment, the evolution of the manganese minerals responded to the increasingly oxidizing conditions, as shown in Fig. 5. First we have the oxidation of the rhodochrosite into hausmannite (6), second the oxidation of hausmannite into bixbyite (7), third the hydration of bixbite into manganite (8), and finally the oxidation of manganite into pyrolusite (9).

(6) $\quad 3MnCO_3 + \frac{1}{2}O_2(g) \longrightarrow Mn_3O_4 + 3CO_2(g)$

(7) $\quad Mn_3O_4 + \frac{1}{2}O_2(g) \longrightarrow 3Mn_2O_3$

(8) $\quad Mn_2O_3 + OH^- + H^+ \longrightarrow 2MnOOH$

(9) $\quad 2MnOOH + \frac{1}{2}O_2(g) \longrightarrow 2MnO_2 + H_2O$

Obduction also provoked the crushing of the rocks and the development of several tectonic systems, the main of which had a NE orientation. Along these fractures we observe dikes of diabase from the Zurrapandilla Complex. During the intrusion of these dikes, the area was affected by a low temperature hydrothermal-metasomatic process. This process developed a well formed listwaenitic zone to which the mineralization is related. The provenance of these fluids is yet to be established, but we can propose three possible origins:

a.- Magmatic origin - orthomagmatic fluids related to the Zurrapandilla magmatic complex.

b.- Slab origin - sea water fluids related to the dehydration zone of the subducted slab.

c.- Mixed origin - fluids that are the result of a combination of the first two options.

Studies done by Ploshko (1963), Zuffardi (1977), Pipino (1980), Buisson and Leblanc (1986), and Pallister et al. (1987) on similar listwaenitic zones concluded that these fluids were composed mainly by H_4SiO_4, H_2CO_3, H_2O, H_2S, K, Na, Rb, and probably $CO_2(g)$ and $CH_4(g)$. Mottl (1991) suggests that these type if fluids usually present high values of pH, high carbonate alkalinity, and low chlorine. In accordance with the mineralogical associations in the area of San José de Las Malezas (Valls, 1995b), I believe that the main metallic component of this fluid should have been copper, with lesser amounts of lead, zinc, silver, and arsenic.

These fluids reactivated the serpentinization of the rocks, so more manganese was released from the remaining olivine crystals. Here, the most probable reaction should have been first- the formation of rhodonite (10), second the oxidation of rhodonite into hausmannite (11), and then the same evolution pattern as shown in equations (7 - 9) to arrive to the formation of pyrolusite.

(10) $\quad\quad\quad Mn_2SiO_4 + H_4SiO_4 \dashrightarrow 2MnSiO_3 + 2H_2O$

(11) $\quad\quad\quad MnSiO_3 + \frac{1}{2}O_2(g) \dashrightarrow Mn_3O_4 + SiO_2$

All these processes are schematically represented in Fig. 5. The formation of these manganese oxides leads to an increase in $fO2$ making difficult the precipitation of gold, copper and other ores in and near the serpentinitic bodies.

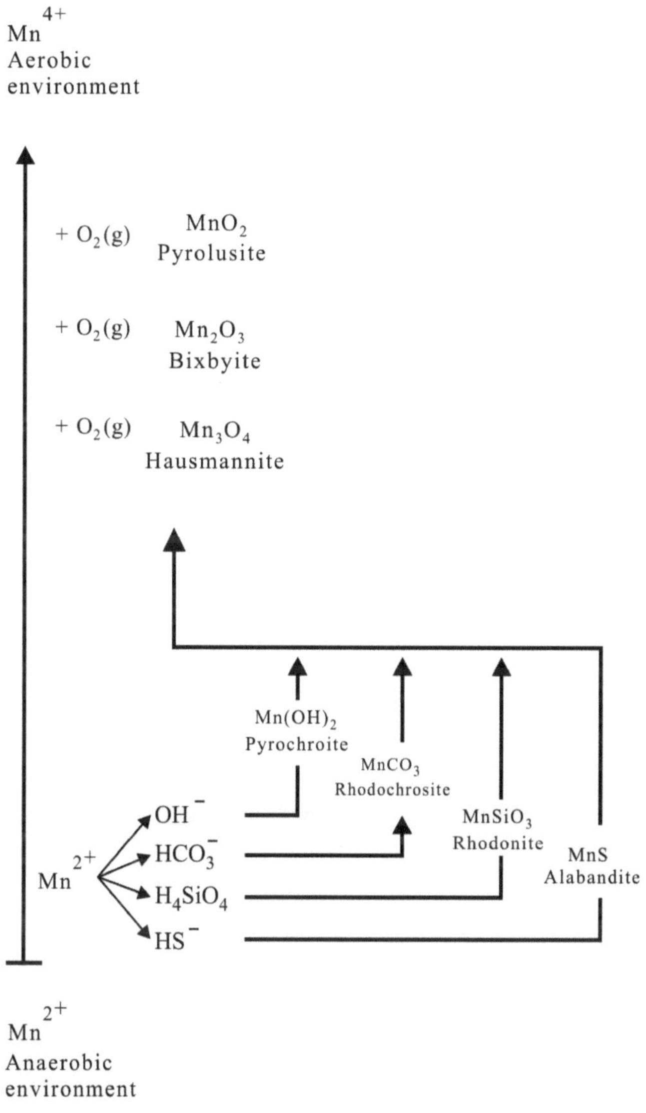

Mn^{4+}
Aerobic environment

$+ O_2(g)$ MnO_2 Pyrolusite

$+ O_2(g)$ Mn_2O_3 Bixbyite

$+ O_2(g)$ Mn_3O_4 Hausmannite

$Mn(OH)_2$ Pyrochroite

$MnCO_3$ Rhodochrosite

$MnSiO_3$ Rhodonite

MnS Alabandite

Mn^{2+} → OH^-, HCO_3^-, H_4SiO_4, HS^-

Mn^{2+}
Anaerobic environment

Figure 5. Evolution of Mn2+ in response to increasingly oxidizing conditions, near 283 °K, and 1 atm.

Dynamic of the Hydrothermal Fluid

Although it has been considered in the past as a typical copper-gold deposit, the correlation study of the existing data from San José de Las Malezas quartz gold vein deposit (Valls, 1995b) shows no significant correlation between gold and copper, and a strong positive correlation between gold and lead in the siliceous zone. Therefore, we conclude that copper and gold are spatially, but not genetically related.

In the proposed model, copper was contributed by the hydrothermal system, with Pb, Zn, As, and Ag, while gold was provided by the serpentinites hosting the quartz bodies. A schematic representation of this process is shown in Fig. 6. Since we already deal with the effect of reactivation of the serpentinization of the host rocks, we will focus on the mechanism of transportation and deposition of the ores.

The listwaenitic alteration consists of four zones, (i) a less altered serpentinites, (ii) an iron-altered zone, (iii) a carbonatic altered zone, and (iv) a siliceous zone (Sawkins, 1990, Valls, 1995b, et al.). These lenses grade laterally into the less altered serpentinites through a talc-carbonated zone.

According to the geochemical results of a meter by meter channel sampling through the four zones (Valls, 1995b), gold, copper, silver, zinc, arsenic and lead were found to concentrate mainly in the siliceous zone. Gold also was found concentrated inside the iron-altered zone, while the carbonate-rich zone is almost barren of ores.

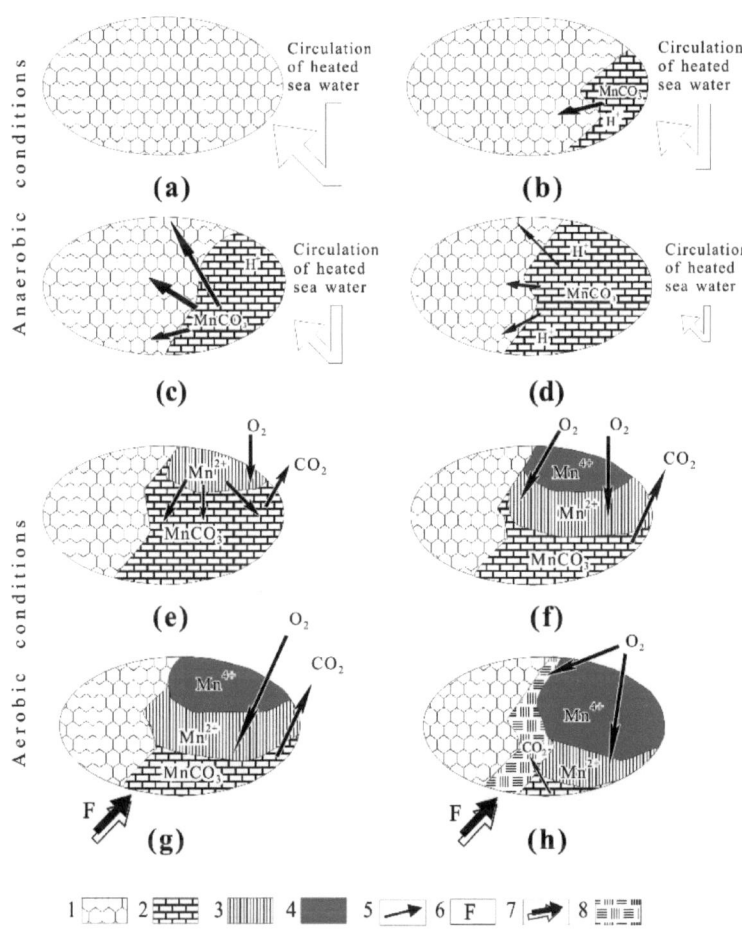

Figure 6. Schematic representation of the alteration of an ultramafic massif and the evolution of the manganese minerals from an anaerobic to an aerobic alteration after the obduction onto the surface of the massif. See text for detailed explanations. 1.- Ultramafic massif, 2.- Formation of rhodochrosite, 3.- Formation of hausmannite, 4.- Formation of manganite, byxbyite and pyrolusite, 5.- Vector of alteration, 6.- Ore fluid, 7.- Vector of mineralization, 8.- Listwaenitic zone.

It is commonly assumed that gold is transported in the (+1) oxidation state (McKibben et al., 1990). Since gold is a soft electron acceptor, it should form especially stable complex with soft ligands as HS^-.

A probable reaction is its transportation as a thio-complex (12).

(12) $\quad Au + 2H_2S + \frac{1}{4}O_2(g) \longrightarrow Au(HS)_2^- + \frac{1}{2}H_2O + H^+$

I also believe that very small and thin scales of native gold could have been mechanically removed by the fluids from the serpentinites. The flat form of these grains allows them to be transported very easily, as seen from our experience during panning to obtain heavy concentrates from these and similar zones. This mechanism of transportation is doubtless less efficient than the one represented earlier (9), but it helps to explain the existence of this kind of native gold scales in the iron-altered zone (Fig. 7).

As the hydrothermal fluid moves toward the surface, several important factors will control its stability.

Inside and near the serpentinites, the evolution of the manganese minerals to their trivalent states, will consume oxygen provoking a reducing environment that difficult the precipitation of gold and other ores.

Closer to the surface, we have first the loss of temperature due to the mixing with meteoric waters, and second, the presence of a more oxidizing environment favoured by the existence of faults and fractures of the rocks.

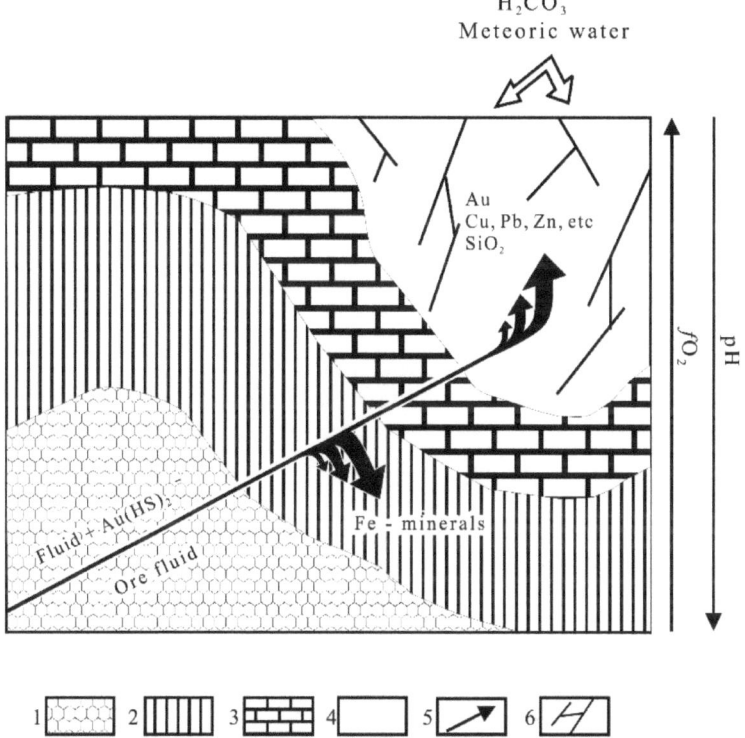

Figure 7. Dynamic of the ore fluid. This model proposes two mechanisms for gold transportation: (i) mechanically -as thin scales of native gold-, and (ii) as a thio-complex. Mechanical transportation explains the presence of scales of native gold in the iron altered zone. Main gold and copper concentration are in the siliceous zone, where the ore fluid got mixed with meteoric waters in a more acidic environment, with low values of fO_2 due to the formation of massicot or galena (see text for further details) 1.- Less altered serpentinites or unaltered ultramafic rocks, 2.- Iron-zone, 3.- Carbonate-zone, 4.- Siliceous zone.

Both the decreases of the pH and the increment of the fO_2, will provoke the precipitation of gold. The precipitation of Au because of a decrement of pH is shown in equation 13 (Spycher and Reed, 1989; McKibben et al., 1990). Equation 14 shows the precipitation of gold due to an increment of the fO_2 (Chris Gammons, personal comunication).

(13) $8Au(HS)_2^- + 6H^+(aq) + 4H_2O(aq) \longrightarrow 8Au(s) + SO^{2-}_4(aq) + 15H_2S(aq)$

(14) $4Au(HS)_2^- + 15O_2(g) + 2H_2O(aq) \longrightarrow 4Au(s) + 8SO^{2-}_4(aq) + 12H^+$

Very often we find strong correlations between gold and lead and we find native gold in galena and lead oxides like massicot (PbO) and crocoite ($PbCrO_4$). This leads us to assume that another possible mechanism of gold precipitation is the formation of galena or massicot as it is represented in equations 15 and 16.

(15) $4Au(HS)_2^- + 4PbCl_2 + 7O_2(g) + 2H_2O(aq)$ --->

$4Au(s) + 4PbS + 8Cl^- + 4SO^{2-}_4(aq) + 12H^+$

(16) $4Au(HS)_2^- + 4PbCl_2 + 15O_2(g) + 6H_2O(aq)$ --->

$4Au(s) + 4PbO + 8Cl^- + 8SO^{2-}_4(aq) + 20H^+$

These reactions could explain the strong positive correlations between gold and lead in the siliceous zone (Valls, 1995b).

Discussion

Is San José de Las Malezas a typical quartz vein copper-gold deposit? The results of the correlation analysis suggest a negative answer to that question, in spite of the opinion of several authors (Arcial and Romero 1985; Cabrera and Tolkunov 1979; Grushdiev, et al. 1966; etc.). I have included here an alternative model of the genesis and mechanism of formation of this deposit.

The model describes the serpentinization of an ultramafic protrusion, which is later obducted onto the surface to form part of an ophiolitic complex. After the obduction, the area was affected by hydrothermal-metasomatic fluids probably related to the Zurrapandilla Suite, along the developed tectonic system. These fluids provoked the formation of a listwaenitic zone toward the less altered serpentinites.

The model assumes that gold was leached from the less altered serpentinites by the fluids, transported as thio-complexes and then precipitated in the siliceous zone due to a decrease in pH provoked by the formation of lead minerals (galena, massicot, etc.), and/or an increment of the fO_2 and the loss of temperature due to mixing of the fluids with meteoric waters in near-surface conditions. The same conditions of pH, fO_2, and loss of temperature in the siliceous zone, provoked the precipitation of copper, lead, zinc, arsenium and silver from the fluid, and the formation of copper and lead-zinc minerals in the contact of the quartz veins with the crushed host rocks.

Conclusions and Recommendations

Based on the information available up to this moment, a model has been presented to explain the geochemical characteristics of the ore distribution in the San José de Las Malezas gold-quartz vein deposit, in Santa Clara, Cuba.

This model describes the serpentinization of an ultramafic protrusion, which is later obducted onto the surface to form part of an ophiolitic complex. Special attention has been given to the formation and evolution of different manganese minerals during the serpentinization of the host rocks. The model helps to explain the negative correlation between gold and manganese over the serpentinites.

It has been suggested elsewhere the possibility of a partial melting process of this ultramafic body before its protrusion. This process could have provoked the formation of a gold enriched zone due to the gravitational separation of this mineral. Although the partial melting of these rocks is possible, this idea needs to be tested in the future. The existence of a gold enriched zone in the serpentinites will not only help to explain the remobilization of gold in the hydrothermal-metasomatic fluid, but also it could become a prospecting objective in the area.

After the obduction, the area was affected by hydrothermal-metasomatic fluids along the developed tectonic system. The origin of these fluids is yet to be determined, but due to the presence of H_2CO_3, H_4SiO_4, and H_2S, it is possible to suggest a magmatic or magmatic-slab origin in preference to a pure slab source.

These fluids reactivated the serpentinization of the host rocks, and provoked the formation of a listwaenitic zone toward the less altered serpentinites.

The model assumes that gold was leached from the serpentinites by the fluids and considers two mechanisms for its transportation. The first one is the mechanical transportation of thin scales of native gold by the fluid. This may explain the existence of a gold enrichment zone in the iron-altered serpentinites.

The second mechanism is the transportation of gold as thio-complexes and its precipitation in the siliceous zone due to a decrease in pH and/or an increment of the fO_2 provoked by the formation of galena, massicot or other lead minerals (crocoite?), and the loss of temperature due to the mixing of the fluids with meteoric waters.

The same conditions of pH, fO_2, and loss of temperature in the siliceous zone, provoked the precipitation of copper, lead, zinc, arsenic and silver from the fluid, and the formation of copper and lead-zinc minerals in the contact of the quartz bodies with the crushed host rocks.

Although this model can presently explain all the known characteristics of the distribution of gold and other ores in this deposit, I am not presenting it as uncontroversial model, but as a working hypothesis to formulate what needs to be tested in future studies.

Acknowledgment

The initial sampling and analytical programmes were carried out by the author and his colleague Eng. Jorge Cruz in 1986, during the preparation of a project for the detail survey of the San José de Las Malezas deposit for the Geological Enterprise "Santa Clara". The Provincial Laboratory associated to the Geological Enterprise provided the analysis of the samples, including the determination of fluorine by a technique designed by the Bulgarian specialist Lazarina Lazarova Ignátova.

I wish to thank Dr. W. Trzcienski for his suggestions and discussions during the preparation of this paper, and also to Dr. Christopher Brooks and Dr. Alex Brown for their review and criticism. I want also to thank Dr. Chris Gammons in particular for his support and guidance, and also for his fair and constructive criticism. Finally, from the big community of the INTERNET, I specially thank Margaret Donelick of the O.D.P. Project and to Dr. P.J. Kenney from the British Geological Survey for their comments and suggestions on the nature and composition of the slabs fluids.

References

Arcial, F. and Romero, O. (1985). Informe de los trabajos complementarios para la búsqueda de oro en el yacimiento San José de Las Malezas. Territorial Geological Library of Las Villas, Cuba, 1985.

Baranova, N. N. and A. B. Koltsov. (1987). The influence of metals and volatiles in hydrothermal solutions on gold transport and deposition based on fluid-inclusions studies. Geochemistry International 24(1): 1.

Baranova, N. N. and B. N. Ryzhenko. (1981). Computer simulation of the Au-Cl-S-Na-H2O system in relation to the transport and deposition of gold in hydrothermal processes. Geochemistry International 18(4): 46.

Buisson, G. and M. Leblanc. (1986). Gold-bearing listwaenites (carbonatized ultramafic rocks) from ophiolite complexes. Metallogeny of basic and ultrabasic rocks. Inst. Min. Metall. 121.

Cabrera, R. and Tolkunov, L.A. (1979). Tipos y condiciones geológicas de localización de los yacimientos de oro de la zona septentrional de la antigua Provincia Las Villas; Ciencias de La Tierra y del Espacio, No. 1, pp 1-29.

Camus, F. (1990). The geology of hydrothermal gold deposits in Chile. Journal of Geochemical Exploration. 36((1-3)): 197.

Clarke, F. W. (1924). The data of geochemistry. USA, Bulletin U.S. Geological Survey.

Crerar, D. A., R. K. Cormick, et al. (1976). Geochemistry of manganese: an overview. In Geology and geochemistry of manganese. 2nd International Symposium on Geology and Geochemistry of manganese., Sidney, Australia.

Cruz, J. and Valls, R. (1989), Proyecto Técnico-Económico Búsqueda Detallada de Oro en San José de Las Malezas. Territorial Geological Library of Las Villas, Cuba.

Farfell, L. S., N. I. Savelyeva, et al. (1984). Hydrothermal solutions at the Asku gold ore deposit. Geochemistry International. 21(1): 71.

Grushdiev, K.I., *et al.* (1966). Trabajos de evaluación, búsqueda y revisión de las manifestaciones de cobre en la antigua Provincia Las Villas. Territorial Geological Library of Las Villas, Cuba.

Hantichak, J. (1963). Trabajos de revisión de las perspectivas cupríferas del yacimiento San José de Las Malezas. Territorial Geological Library of Las Villas, Cuba.

Kanchev, I.L. et al. (1973). Informe del levantamiento geológico 1:250 000 de Cuba Central. Territorial Geological Library of Las Villas, Cuba.

Lofgren, G. (1980). Experimental studies on the dynamic crystallization of silicate melts; In Physics of magmatic Processes, pp. 487-551.

McKibben, M. A., A. E. Williams, et al. (1990). Solubility and transport of platinum-group elements and Au in saline hydrothermal fluids: constrains from geothermal brine data. Economic Geology 85(8): 1926.

Mottl, M.J. (1991). Pore waters from serpentinite seamounts in the Mariana and Izu-bonin forearcs, leg 125: Evidence for volatiles from the subducting slab, in Proceedings of the Ocean Drilling Program. Scientific results. V 125, 373-385.

Pallister, J.S., J.S. Stacey et al. (1987). Arabian shield ophiolites and late Proterozoic microplate accretion. Geology (15): 320.

Pipino, G. (1979). Gold in Liqurian Ophiolites (Italy). International Ophiolite Symposium, Cyprus.

Ploshko, V. V. (1963). Listwaenitization and carbonation at terminal stages of Urushten Igneous Complex, North Caucasus. International Geological Review (7): 446.

Rodriguez, C. and A. J. Warden. (1993). Overview of some Colombian gold deposits and their development potential. Mineralium Deposita 28(1): 47.

Sawkins, F. J. (1990). Metal Deposita in relation to plate tectonic. New York, Springer-Verlag, Berlin-Heidelberg.

Spycher, N. F. and M. H. Reed. (1989). Evolution of a Broadlands type Epithermal ore fluid along alternative P-T paths. Implications for the transport and deposition of base, precious and volatile metals. Economic Geology 84(2): 328.

Valls, R. A. and I. Gonzalez. (1987). Evaluación geomatemática de los muestreos litogeoquímicos en mina Descanso, Villa Clara, Cuba. Moa's Mining Metallurgical Institute.

Valls, R.A. and Nuñez, F. (1988). El uso de los métodos geoquímicos para la búsqueda de yacimientos no metálicos en el Macizo Metamórfico del Escambray; Revista Tecnológica, No. 1, pp. 12-29.

Valls, R.A. (1995a). Geochemical peculiarities of the quartz-gold deposit of San José de Las Malezas, Santa Clara, Villa Clara, Cuba; Term paper, Université de Montréal.

Valls, R.A. (1995b). The role of manganese as a controller for gold mineralization in the serpentinites of the San de Las Malezas quartz-gold deposit in Santa Clara, Villa Clara, Cuba; Term paper, Université de Montréal.

Yardley, B.W.D. (1989). An Introduction to Metamorphic petrology; Longman Scientific & Technical, New York.

Zuffardi, P. (1977). Ore/mineral deposits related to the Mesozoic ophiolites in Italy. New York, Springer-Verlag, Berlin-Heidelberg.

About the Author

As a professional geologist with thirty-two years in the mining industry, I have extensive geological, geochemical, and mining experience, managerial skills, and a solid background in research techniques, and training of technical personnel. I am fluent in English, French, Spanish, and Russian. I have been involved in various projects world-wide (Canada, Africa, Russia, Indonesia, the Caribbean and Central and South America). Projects included from regional reconnaissance to local mapping, diamond drilling and RC-drilling programs, open pit and underground mapping and sampling, geochemical sampling and interpretation, and several exploration techniques pertaining to the search for diamonds, PGM, gold, nickel, silver, base metals, industrial minerals, oil & gas, and other magmatic, hydrothermal, porphyritic, VMS and SEDEX ore deposits. Special strengths are related to acquisition of new properties, geochemical and geological studies, management and organization, geomathematical analysis and modelling, compositional data analysis, structural studies, database design, QA&QC studies, exploration studies and writing technical reports. P.Geo. registered in the province of Ontario.

www.ingramcontent.com/pod-product-compliance
Lightning Source LLC
Chambersburg PA
CBHW041610180526
45159CB00002BC/796